SAFE
DRIVING

Other Titles by Lloyd Cole

Let Me Die
A New Life of Jesus
In Defence of Private Enterprise
My Baby, My Body, My Choice
The Philosophy of George Bernard Shaw
An Introduction to Poetry
Successful Selling
Mr Churchill and the Church
Modern Miracles
The Chinese Prince
The Windsor Story
Ye Gods

Compiler of

The Solvit Crossword Puzzle Series
What Do You Know? Crossword Puzzles

SAFE DRIVING

by

Lloyd Cole

1992

Lloyd Cole

MAIDENHEAD

First Published 1992

Lloyd Cole
37 College Avenue
Maidenhead, SL6 6AZ, UK

© Lloyd Cole, 1992

ISBN   1 874052 21 2 (hbk)
1 874052 22 0 (pbk)

A record of the Cataloguing in Publication data
is available from the British Library.

Printed in Great Britain
by Cromwell Press

CONTENTS

continued

PREFACE

Today is the age of the specialist. Given a problem, the thing to do is to call in the expert.

Who is the specialist with regard to driving a car? Who is the expert? Some of the top racing drivers get involved in accidents on the roads and are judged to be in the wrong, and are fined or suspended from driving on public roads. One at least died as a result of an accident on the road which could not be blamed on any other driver. Who then is the expert? It appears that the expert racing driver may not be the safest driver on the roads. To whom then can we turn for advice on safe driving on the road? Obviously, many racing drivers are very good drivers under all conditions, and much that a racing driver learns must help to make him a competent driver in all spheres of driving.

However, with reference to driving on the roads, I think that for the best advice on driving under everyday driving conditions, we should turn to a man whose experience on the road has been considerable and blameless.

All the above is said because I am suffering from something like an inferiority complex because I am not a Stirling Moss or a Nigel Mansell and feel quite sure I never could be. People, before reading a book on driving, are apt to ask concerning the author, "Who is he? What has he done? Is he an expert? What races has he won?" I am not a racing driver. I have no idea if I could be a good racing driver or not. I have no claims in this direction at all. I can only say that I have driven a hell of a lot of miles under all kinds of conditions; that I still do so; that in general I do travel faster on the roads than other drivers, and that despite doing more driving and driving faster, I do not have serious accidents. This is my only reason for writing this book — the thought that perhaps from my experience I can make suggestions or throw out hints that will result in fewer people being injured, killed, or bereaved because of accidents on the road. Even more than this perhaps — that many who drive to get from here to there may begin to enjoy their driving for the sheer pleasure of it. This gives me two reasons for writing, **to save lives and to promote pleasure.** Do not smile at this because it is true that real driving is a pleasure. The man who has mastered the art of driving so that he can drive to the limits of his machine without danger to others is a happy man indeed.

The only worth-while test of good driving is on the roads. To take a corner fast and safe on a racing track with no traffic coming the other way is one thing. To learn to corner fast and safely on all kinds of roads with traffic coming the other way is another thing altogether. In both cases there is a limit beyond which there is no return. Step over this limit on the roads, and the least that can be said is that you are acting like a would-be suicide and the worst that you are acting like a would-be murderer.

In the pages that follow I shall suggest some things that all will not accept. I will suggest that only one thing causes accidents — damn bad driving. No two drivers who are driving correctly can possibly have a head-on collision. Given quite normal road conditions, no driver can leave the road on a bend if he is driving within the limits of his car and his ability. Overtaking is never dangerous to a motorist who is observing the necessary precautions. No man runs into the car in front of him when it stops for an emergency if he is driving correctly.

I will suggest that **stupidity causes more accidents than speed,** and that a car is the only lethal weapon which is put indiscriminately into the hands of all and sundry without proper training.

If only because of the appalling slaughter on the roads, every car owner in the country should determine that he or she will learn to be efficient in the use of a car under all conditions. However, it need not be so only for this reason — it can also be for the joy and pleasure of accomplishment. By learning to drive safely you can avoid increasing the grief in the world while at the same time greatly adding to your own pleasure.

1   REASONS AND EXCUSES FOR ACCIDENTS

Cause and effect is a well established principle. If a pot of paint falls on a man walking under a ladder it would be reasonable to expect that a painter had left a can of paint insecure. If two cars collide in the street there is always a reason. Accidents of this kind do not just happen. Excuses like, "There is such a lot of traffic on the roads, I wonder there are not more accidents," are not acceptable. So there is a lot of traffic. That makes it easier to see. It's constant and intrusive presence should lead to the extra caution needed to avoid collisions.

Another and very popular excuse is that how ever careful you are, you may have an accident because of what other people do. The other fellow is to blame. You may drive very well but some idiot will involve you in trouble. I am not prepared to accept this sort of reasoning. Apart from anything else surely it must be agreed that if we are to believe that every time we go out driving there are many idiots driving badly and by so doing preparing conditions which may lead to our being involved in an accident, then surely we shall take special steps to avoid becoming involved in such accidents. Forewarned is forearmed. If, in spite of the knowledge of what to expect one is still involved in an accident then some of the fault at least must attach.

Supposing you go out at night to walk across a field. You are warned of certain pitfalls. Advised to carry a stick and a torch. Told that there is considerable danger of accident if you do not observe these simple precautions. You start out as advised, but after a while you think the precautions unnecessary, and throw away either the stick or the torch or both. You fall into a pit and break a leg. Was this accident entirely the result of the conditions in the field? Or was it at least in part the fault of someone who did not take the

1

necessary precautions to avoid an accident? Could the accident have happened if the person concerned had been properly equipped to cover the terrain being travelled? If you learn how to travel properly in the conditions you have to face there will not be an accident. Exactly the same applies on the road in a great many cases. If I may be forgiven for quoting my own case, I travel about 1000 miles every week and I am therefore more exposed to accidents than the majority. Over and over again things happen which could mean an accident. People pull out of roads right in my path. People pull out of a line of traffic as I am passing. Other drivers stop suddenly in front of me, turn right or left without signal. All these things happen to me, and several times a day I could be involved in an accident. Yet I am not so involved.

If I was involved, I would find it difficult to believe that the accident could not have been avoided. I agree, of course, that many accidents are blatantly caused by one person, and that the others involved have no blame as far as the cause of the accident is concerned, but it does not necessarily follow that they could not have avoided being involved if their own reactions and preparedness had been more professional, and closer to perfection.

The pitfalls in the field of our illustration were real enough. Perhaps they should not have been there. However they could have been avoided quite easily by taking certain precautions.

The other man may be the cause of an accident, but does this always excuse you for being involved? Supposing you had braked sooner, or steered clear? Suppose you had being paying attention to your driving instead of woolgathering?

The police must hear the same things all the time. Such as, "I didn't see him. Why he came straight at me," or, "I didn't have a chance." And so on.

Let us agree that you may be involved in an accident which is not your fault. Consider only this — need you be involved? Unless there is absolutely no way out I will not be involved in an accident. An accident will never happen to me because my reactions are slow or because I have not anticipated a situation which could have been foreseen. Can you say the same? If not, then you are partly to blame for any accident involving a car which you are driving at the time.

Read the chapters that follow and learn to **really** drive. Make certain you are properly equipped for the journeys you intend to make. Do not think that passing a driving test makes you a driver. It does nothing of the sort.

Firstly, you must not be one of the persons to cause an accident. But this is not sufficient. You must learn enough to be able to see to it that as far as is humanly possible you do not become involved in accidents caused by others. Remember the field. The pitfalls are there, but you can be equipped to avoid them.

It is a favourite saying of many people that it takes two to make an argument. It is equally true that it takes two to have a collision. You must drive so well that you are **neither** of them.

2   STUDY TO DRIVE WELL

It has been said many times in connection with all sorts of occupations and pursuits that nothing really worthwhile was ever accomplished without enthusiasm. This is just as true when applied to driving. If you have bought a car just to get from place to place, and you have no real enthusiasm for driving, may I recommend that you sell it. There are far too many people on the roads who have no real interest in driving as an accomplishment.

Such people are a menace.

If you must own a car let it be that you really want to drive. People collect stamps or butterflies with an almost fanatical enthusiasm. Yet, quite often, these same people will drive a car without any thorough study of driving. The fact that they have passed a driving test when on their best behaviour is all they worry about.

I have talked to some very intelligent people about driving. These are people who study every minute they have to improve their knowledge of everything to do with their profession. Yet, when it comes to driving a car they learn the rudiments and leave it at that. Some of the most intelligent professional men are the worst drivers in the world.

The most important things are often the most neglected. Many a man will work himself half to death building a business while he neglects and finally loses his wife. Another person will put every spare minute into studying to learn how to be a good solicitor or a doctor, and then get themselves killed in their car on the way home because they never bothered to learn to drive really well.

Can anybody explain to me why so few car owners become good drivers? Is it true that the average man does not care whether or not he learns to drive competently? There are too many people being killed and maimed on the road to permit anybody to continue driving badly. The person who will not take the trouble to learn to handle a car safely is at least anti-social and possibly a potential murderer. Let us not mince words. A man who could drive safely if he would only put his mind to it, but does not do so, and injures or kills someone as a result of his amateur and sloppy handling of a car, should be banned from driving for life regardless of whether this is a hardship or not. Such punishment is little enough for a man who will not be socially decent enough to behave with some regard for the safety and well-being of others.

Some time ago I was driving in London in the evening rush hour. Now, I am a fast driver, really fast. I learned to handle a car and drive at speeds between 80 and 105 miles an hour on all kinds of roads all over the country, and at 120 miles an hour occasionally on a few roads when there was no 70 mph limit. Nevertheless, I have never seen such dangerous driving as I saw on this road in ten minutes. I was really afraid that maybe I was not going to get home without my first serious accident. On this particular evening it was dusk, which is the most dangerous time of the day for driving. In addition it was raining hard and the roads were slippery. Yet, a whole lot of drivers were so intent on getting home first that they threw all caution to the winds, dropped all courtesy, and drove like raving lunatics. **This is not fast driving, it is sheer lunacy.** I must admit I do not have the nerve for it, yet I will wager that none of the maniacs I saw that night would last ten minutes in a test of true driving skill.

The first law for safe driving is enthusiasm, by which I mean a genuine desire to **learn to drive as an art.** Real motoring is an art. In all the arts the aim is perfection. It is so with driving. Obviously, only some artists reach the top of their profession but most will try with enthusiasm to do so. It should be so with driving. Sloppy and untidy driving is the curse of the roads today.

Because the accident rate continues to mount a search is made for scapegoats. The roads are blamed. The large number of cars is blamed. The easiness of the driving test is blamed. The fact is that only one thing is to blame — bad and inefficient driving.

It is the duty of every car driver to study to become an artist of the road. If only out of social conscience this is so. It is worth while to do things well. Also it is most rewarding. There is very little that offers as great a thrill or pleasure as complete mastery of a car.

Have the will to be a good and safe fast driver. Study your car. Learn what can be done with your car. Take time off to test your car at different speeds under different conditions. Try a bend over and over again until you know how to take a bend in the correct line. Then find out how fast you can do this safely. Learn to use your steering instead of being half afraid of it. Do not be afraid of a machine. Man was not made for the machine. Believe that your car will only do what you tell it to do. Learn what you should tell it to do. Think no longer of a car as a means of conveyance. A car is an adventure. Regard it as an adventure. To enjoy the adventure it is essential that you should study every aspect of driving. In the pages that follow I shall endeavour to point out what constitutes bad driving, and to indicate as far as possible the way to competent and efficient driving. Many driving mistakes are habit, or are not recognised as mistakes because no real thought has been put into the question of perfecting driving technique.

In business, the successful businessman is the one who considers all sides of every question of policy, and finding the right course of action, follows it without compromise until such time as experience may teach that error has crept into his calculations. Then, he does not decide that he is too good a businessman to have been wrong. He knows he was wrong and sets about correcting his mistakes before these lead to financial disaster. He is willing to go on learning all the time, and thus perfecting his business technique.

A similar study of the art of driving is essential. Also a willingness to learn, but in the case of driving it is better to learn other than by experience as the experience may mean death on the roads. For this reason it is hoped that all car owners will be willing to learn all they can from this book, so that on the roads they may be better drivers, and fewer families will be bereaved as a result of senseless carnage on the road.

The successful businessman reaps the reward of his prudence, his willingness to learn. So with the driver. The exhilaration that comes to the person behind the wheel who is master of their machine is most rewarding.

Read therefore, and learn, and put into practice.

3 LEARNING TO KNOW YOUR CAR

All cars are not the same.

I have read several times in the papers of someone taking out a new Jaguar for the first time and being killed. It may well be that these same people were perfectly safe drivers in another make of car, less powerful perhaps. People read of the methods of racing drivers and like to copy. This is quite laudable providing it is remembered that it takes a little time to become a top-class racing driver. Nobody should expect to corner at high speeds until they have learnt by degrees the speed at which they can control their cars in the line they want to follow round a corner.

Many accidents occur to drivers of fairly new cars. They have probably driven quite safely for years in other cars. This leads me to say that it is important to learn to know your car.

It is a good idea to take time off to experiment with your car. Choose a spot and a time when the traffic is as light as possible and find out what your car will do under different circumstances. Anyone can corner at 20 miles an hour, although I watch some people use nearly all the road to take a bend even at this low speed. However, for the purpose of this chapter we will presume that at low speeds you find it possible to follow a line through a bend with perfect safety and near perfection of style.

Now, you want to be a good driver. To be able to drive faster without any loss of control, and without peril to yourself or other road users. Why not practise to this end. Most perfections are attained by trial and error. Providing you choose the right place and the right time you should be able to learn quite a deal about the way your car will respond at different speeds and on different surfaces.

I must say here that the performance of a car will not be good if the car is not maintained in good condition. For instance, I am presuming you will never drive with worn tyres, and that your brakes and steering are always maintained in tip-top condition and adjustment. Granted that this is the case you can soon learn the limits of your car and your driving ability. Having done so, drive within those limits, and remember that in bad weather the limits should be lowered in your mind by at least 20 per cent.

Having selected the right spot, experiment with different speeds and different methods on bends. Make your limit for future driving just below the top speed at which you are still in perfect control of your car. I say, just below, because you should always have some speed in hand for an emergency. I have found my way out of trouble more often by an increase in speed than by braking. You may regret very much finding yourself in a situation where you have no reserve of power.

Experiments with your brakes will show you quite definitely the distance you need at all speeds to pull up safely. Remember that you may well need a little more distance in an emergency than you need when you know you are going to stop. Find out if your car has a tendency to pull to either side and allow for this in your driving if you are unable to have it corrected. Remember any such tendency will increase on a wet road.

Make a note of car-reaction to braking and steering on different road surfaces, and apply this knowledge when driving normally. Do not expect your car to perform exactly the same at the same speed if the road surface is different, because it will not do so.

I know all this sounds very elementary, yet how many people bother to undertake any such experiments to learn how their car will behave?

Have you ever seen the look of surprise on a driver's face when he brakes in the middle of a bend and his car goes into a slide? Why the surprise? Unless there is some mechanical defect in a car the good driver can never be surprised at what happens in a given set of circumstances. That is why the good driver usually comes out of a skid as quickly as he goes into one. He knows what to expect under certain prevailing conditions and he knows what to do about it.

You must not expect to be able to do on the roads all that you see done on the racing track. It is perfectly possible to induce a slide so that you can take a sharp bend at 70 miles an hour with absolute safety on a racing circuit, but not on a busy road. If you are competing in a race on a circuit you may well accelerate to overtake on a wet road surface at over 100 miles an hour despite an approaching bend. You may dice with death because this is your profession. On the roads however, you must not dice with death under any circumstances. Not only your life is at stake, and the only prize you are likely to win is the loss of your no-claim bonus, and, possibly, seven years in jail for manslaughter.

Just as you learn to use the tools of your trade you must learn to use your car. A car is a machine, and properly used and maintained it will serve you well. Especially, if you earn your living by travelling, it is important that your mastery of your car under all circumstances be of the very highest standard. Why do you think insurance companies charge higher premiums to commercial travellers? Because they drive more and are therefore more exposed to danger. It is no good being a good salesperson if you end up a dead one.

4   SPACE ON THE ROAD

Sloppy drivers take up far more than their share of the road, and one of the signs by which you can recognize a good driver is his **correct positioning** in the road, and the fact that he does not hinder the progress of other cars by occupying more space than is necessary.

When observing other drivers in front of me on the road I am constantly astonished at the fact that the great majority have no idea how much space there is on their near side. It seems that they must feel that the near side of the road is electrified, and that they must not approach anywhere near it.

I have even seen drivers involved in a collision with a car coming toward them when they have more than two metres of space on their near side. The man coming the other way may have been overtaking at a point when it was risky to do so, but it was made risky only because the driver approaching him had chosen to drive nearly in the middle of the road. These are the road-hogs, they hog the road.

On a road which is wide enough for three cars abreast it should be possible to overtake at any time. This is made impossible or dangerous by those drivers who will not keep in to their near side. Many a time I have to wait to overtake for this reason. This leads to accidents, because motorists get impatient if they know there is room to overtake, and they are being prevented from doing so unnecessarily.

That is why I contend that the sloppy driver who wouldn't dream of driving faster than 45 miles an hour is very often the cause of accidents. Yet these people are praised by Ministers of Transport for driving slowly. The truth is that they are the biggest curse on the road.

These "keep well out in the road" slouches should learn to use the roads properly. Properly means bearing in mind the fact that there are others on the road, and that some of them may wish to travel faster for reasons of business or emergency. **The good driver is never the cause of holding up other motorists unnecessarily.** Never.

There is far too much accent on "speed" as the cause when accidents increase. I contend without the slightest doubt in my mind that the efficient fast driver is the least likely of all to cause or be involved in accidents. I have proven this in my own driving.

I should like to see the emphasis changed from the call to drive slowly to a call to drive correctly or efficiently. Speed limits will never reduce accidents while sloppy inefficient driving continues. If all drivers would drive correctly speed limits would not be needed. To an efficient driver they are not needed now. I do not need speed limits to tell me when to drive more slowly. On the other hand the idiot driver of today, outside a speed limit, drives at any speed he likes without a thought of the conditions prevailing. There are areas where speed limits should be imposed and others where they should not be imposed. All areas call for a correct driving technique in accordance with the car being driven, the abilities of the driver, and the condition of the road as to weather and volume of traffic.

One of the first things to learn if you would achieve this efficiency is the correct use of the road. If you are travelling one way and another car is travelling toward you, while yet another wishes to overtake the first approaching you, do you drive so that this is possible, or are you so far out from your near side that the overtaking motorist is held up until you pass?

As a fast driver I am held up many times every day because other motorists are either too selfish or too stupid to pull in to their near side in circumstances where conditions demand this to enable traffic to flow smoothly.

The motorist who could never drive fast with any degree of safety is always up in arms against the fast and efficient motorist. He asks why should he not amble along if he wants to? What these people do not realise is this. No fast driver will object to slow traffic which behaves properly on the road. A slow driver should keep over to his near side. He might well make it a point of pride that he drives so that nobody is ever held up unnecessarily by him. He should open his driving window – not sit huddled up in an airless car – so that he can give helpful signals to faster traffic, thus helping the fast man on his way, reducing the possibility of accidents, and at the same time finding more pleasure in his own ambling in so far as he does not have cars on his tail all the time trying to pass.

Whether driving fast or slow, every driver should regard it as part of driving to help every other driver. Why not? Why should the man who prides himself on his manners act like an inconsiderate lout once he is at the wheel of a car? If a person will give up their seat to someone on a bus, or open a door for them, why should they not carry these manners into their driving?

There really is no sensible reason why normally decent people should become completely devoid of all consideration when driving a car. Neither is there any understandable reason why normally intelligent people should drive their cars as though they were born without a single brain cell in their heads.

5   DRIVING SCHOOLS

If you have a member of your family who wants to learn to drive, please have them taught by an efficient driving teacher and not by a member of the family or a friend. Let it be said quite plainly that a husband is not usually the best teacher for his wife in this matter, and vice versa.

On the other hand some driving school teachers do not appear to do their job very thoroughly. I have driven behind cars with driving school instructors at work teaching someone to drive, and have been astonished at the apparent lack of correct instruction. I presume that by the time a driving instructor brings a pupil out on to main roads he has taught his pupil the rudiments at least. Yet I have been behind driving school cars with an indicator light left on for a considerable distance after a turn has been made and while passing several more turns.

I have been behind other driving school cars in which the instructor seems quite happy to let the learner crawl along in the middle of the road. I deliberately stayed behind one longer than necessary for the purpose of discovering if the instructor would do anything about this glaring fault. But not on your life. Not only was the pupil permitted to carry on for over a mile in the middle of the road but was permitted to swing out even further to make a **left** turn! No wonder some people do this all their lives, acting always as though they had an articulated lorry instead of a small car, and in the process holding up other traffic.

It is most important that the people who teach other people to drive should be masters of the art they teach. Yet a great many of them are not. I have seen some of them driving the instruction cars on their own, and their lack of positioning or road sense show only too plainly that they have never taken the trouble to perfect their own driving.

This is an appalling state of affairs, and something should be done to ensure that all driving instructors are fit to teach others to drive.

When any member of your family wishes to learn to drive, please make all the enquiries you can to ensure that he or she is properly taught. The letting on to the road of people who cannot drive correctly, and have not even been taught to do so, must stop.

6   STARTING

At first sight it does not seem that much can be said about starting a car, but there are some things that should be said.

One thing to do every morning when you start your car is to check brakes to be sure these are efficient and there has not been a leak of fluid during the night. Drive a few yards and stop. Once satisfied that brakes are in good condition move off. With many people this check is automatic when backing out of a garage or drive, but if this is not so, do not move off at a fast pace up the road without this check. You may go all your life without brake failure, but brake failure does occur, and it is better to know about it when it does.

Never pull out to start off without making sure there is nothing coming. If you move off when something is coming at some distance then for heavens sake move off sharply. Do not amble out so that you hold other people up, and then act surprised if a toot on the horn expresses the other driver's displeasure. The driver has every reason to be displeased. If you are one of those drivers who makes a meal of moving off, and a life's work of changing up through the gears, then do not move off until you will not be a nuisance to people who can drive.

To revert for a moment to the question of making sure that nothing is coming before moving off. A glance in the driving mirror is not always enough to ensure this. It may well be necessary to turn and look behind or a blind spot on your car may result in yet another cyclist being knocked off their bicycle by a driver pulling out.

Better safe than sorry.

Quite a considerable number of accidents are caused by motorists pulling out from a line of parked cars into the path of oncoming traffic. That one should never do this is surely obvious even to a moron. Yet accidents are constantly caused by this very imbecility.

If you are in a hurry, or a business call has not worked out as you would wish, this is not sufficient cause for dangerous driving. Once you enter the driving seat your mind should be concentrated on the job of driving correctly and safely. Absentmindedness is unforgivable in a driver. Lack of attention is a criminal act and deserves punishment as severe as possible.

All acts of drivers which are not executed in the interest of other road users are to be condemned.

We live in an age of hustle and bustle, but if this state of mind enters into our driving it is doubtful if we will live very long at all. A driver must be calm and in control of himself. His mind must be free of all distractions which will cause him to behave in any way which becomes dangerous in the driving seat of a car.

Most accidents could be avoided. All accidents resulting from a car moving out into traffic could be avoided. There is absolutely no excuse for accidents of this kind. They are caused by downright carelessness, and a careless driver is a terrible thing.

The time taken in moving out is of some importance also. Sometimes a driver makes the mistake of moving out in front of me. He sees his mistake too late, but does he hurry out and away so as not to impede my progress. Very seldom. Usually he moves out as slowly as can be imagined. Then he takes ten times as long as necessary to change gear and get on his way, thereby increasing the time spent obstructing.

All motorists moving out into traffic should move, not dawdle.

Also, in moving out into traffic do not complete a full half circle in the road. It is not necessary to do so. **Steering wheels are meant to be used** so that you can keep in as you pull out. There is no need to create a break in the smooth flow of other traffic, and a good and thoughtful driver never does so.

Pull out therefore, only when you are absolutely certain it is safe to do so, and if traffic is approaching in the distance, get a move on. Tomorrow will not do. Do not create a situation where a motorist coming along has to pull out round you so far that he may be involved in a collision with something coming in the opposite direction.

Pull out when safe. Pull out speedily. Keep in the space you really need.

7 WHERE TO DRIVE

To digress for a moment from instructional matters, may we consider another aspect of modern motoring. Namely, the volume of traffic on the roads. Even at a weekend, when the purpose is pleasure, the main roads are packed with cars. The byroads are more or less empty. This is a most peculiar state of affairs to say the least.

There is very little pleasure to be found on main roads. Even in beautiful countryside the best views are not usually obtainable from the main roads. How on earth people can regard it as an outing in the summer to crawl along main roads for 20 to 50 miles, then park on the verge and eat a picnic with cars whizzing by all the time, finally to crawl home through another traffic jam, I do not know. **It seems to me that we are becoming fascinated by our own stupidity to such an extent that not only do we take part in stupid acts, but we like to sit and watch others doing the same.**

I live near to much beautiful countryside, and on a Sunday the main roads are crowded with cars. After driving a short distance along one of these I turn off and follow any small interesting looking road or lane. Imagine my surprise that during an afternoon of wonderful pleasure from views enjoyed, woods wandered in, a picnic in the fresh air in the heart of the country, I find that I have hardly met a single other car. Back on the main roads, the drivers are stolidly plodding on looking at each other's bumpers, or pulling in to the side, and eating amid the stink of exhaust fumes. Is there any escape from the point of view that most people are raving mad?

On the way home these main road motorists are all a bit weary, fed up with the nervous strain of convoy driving, and several accidents occur each weekend as a result of the frustration and impatience thus engendered. Yet the next weekend sees exactly the same thing repeated again with many of the same characters.

In this book may I make a plea for sanity, even if only because this will mean fewer accidents? Why not get off the main roads when seeking pleasure? You will find much more if you do. The little lanes and byroads in the country lead through some very pretty villages; to some very old and interesting churches; to the best vantage points for a good view of the countryside; to fresh and healthy air and clean and unspoilt countryside for picnicking; to many splendid woods which can barely be seen from the main road. Also, if you like a drink, where better than a nice cosy little country pub away from the hurly-burly of the crowds.

The city dwellers, suffering a life of strain all the week will be rejuvenated as a result of a visit to the country where instead of fighting through heavy traffic, they can relax driving in near solitude, stopping to enjoy the beauties of the countryside without holding up a lot of angry impatient people in other cars.

Also, the likelihood of being involved in an accident is considerably reduced. Firstly, because the traffic is less, and secondly, because, even when they rejoin the traffic they will be fresh and refreshed from their outing, and their own driving will be the better for it.

8 POSITION

This chapter, will of necessity, cover some of the same ground as the chapter dealing with space on the road, but there are one or two more things to be said about position which were not covered in this earlier chapter.

It is unskilled driving, not the number of cars, which causes congestion on the roads. Sloppy and untidy driving, usually at a snail's pace, is the cause of the big traffic jams which occur all too often.

First, consider position when stopping either at stop signs, traffic lights, or to park. Each half of a great many roads will accommodate two cars side by side, but what do we find quite often when traffic comes to a halt at a road junction or the lights? One motorist after another stops dead in the middle of the space available. Why? Where there is room for two cars there should be two cars when the roads are busy, and any driver who hogs the middle of the road is guilty of careless driving, and is increasing the tension in drivers behind, some of whom may have appointments to keep. This leads to impatience which in turn produces the tendency to take risks and often results in an accident. The cause of it all is probably still dawdling along in the middle of the road at his "safe" speed. Safe. This driver is the least safe on the road if we are considering the safety of all.

If you would drive competently, **drive to position.** When you stop, stop in position. If you are turning left or right stop pulled in to the left or over to the right. You do not need the whole road to turn in, or if you do, then you should not be allowed to drive a car.

While considering position when stopping we must speak of parking. Parking is fast becoming the national sin. More people pay fines for parking every year, yet these fines are usually for parking where no obstruction could possibly be imagined as caused, and is the cause of much ill-will between the motorist and traffic wardens.

This kind of parking has not really any connection with safe driving, but perhaps a word or two on the subject will not come amiss in a book intended for motorists.

Parking in towns and cities must be permitted to businessmen and especially commercial travellers with heavy bags of samples and demonstration pieces. That such people should be fined for doing what is essential to their livelihood is monstrous. Provision should be made for such people to carry on their trade.

With reference to others who park because they want to get a cup of coffee or do some shopping. These are better in a car park and not on the roads in busy thoroughfares.

Parking meters are a blessing in one sense, namely that more people have an opportunity of parking near to somewhere they have to call. Why they should have to pay to do so I do not know however. The motorist pays considerable taxes for the privilege of motoring and has never received value for money.

However, these matters are outside our subject and the above few reflections will have to suffice. To consider the aspect of parking that is concerned with safe driving however brings up quite a number of points for discussion. The point that worries me most is that the parking which is most likely to cause accidents usually goes unpunished.

As I travel a great deal I am constantly coming up against the most idiotic and dangerous parking imaginable. On fast main roads cars are left on hills and on bends constantly. I think they must be parked by people who think nobody ever drives faster than themselves and that therefore they will not cause much trouble. I wish people would realise that it is quite possible that a motorist will round a bend at well over 50 miles an hour, and that a car parked just round the corner is very likely to cause a serious accident if the road is not too wide and another car is coming the other way. In fact, a serious accident could be caused if both moving cars were only travelling at 30 miles an hour. If you leave your car in any place where it may be the cause of an accident you are guilty and fully responsible for the accident that occurs. This is obstruction.

There are people who, out to enjoy a country jaunt, think it quite in order to pull in and park where the view is nice. The fact that their car causes several hundred others to have to stop, and may cause an accident quite easily does not concern them in the least. Anyone who parks their car in such circumstances has no conception of correct and efficient driving. A person who has learned to be an efficient driver and is therefore of necessity a fairly fast driver would never park a car in any position likely to cause danger. The fact that someone does so park disqualifies them from any claim to be good drivers.

Having dealt with position when stopping we must consider position when moving along. Good driving is smooth as well as fast. You cannot drive smoothly if you are never in the right position for what you intend to do next. The driver behind you will have great difficulty also if you drive without consideration of position. We have said in an earlier chapter that only sufficient space should be occupied on the road. It must now be added that it is little use keeping to the correct amount of space if it is the wrong space. In other words, if your driving is to be correct it must be smooth, and to be smooth you must be moving always into the right position to achieve your driving purpose without jerkiness. The unskilled driver is always being caught off balance. He never knows until the last minute what the motorist in front is going to do, and consequently is never ready to do what he needs to do.

Position is important. When I am behind a skilled driver on the road it is really unnecessary for him to give me any signals as I can see from his positioning when he is going to overtake, when he is going to turn left or right. It is perfectly obvious. But only one skilled driver recognises the reasons for the moves of another. I remember being behind a "middle of the road, want to overtake, but never will" type of driver who was following a driver who was obviously much more competent. The frontmost driver was obviously moving into position for a right turn in the near future, which he did signal in good time. As he began to move into position, I, two cars behind, began to move in to my near side to be in position to pass when he made his turn. Here are two cars positioning themselves to be able to continue a smooth journey without interruption. Not so the car immediately in front of me. He stuck out in the middle, and close behind the man in front. Result — when the first driver began to make his turn the second saw the signal at the last minute, had an awful job to avoid running into the turning car, violently veered to the left, and had I not been aware of what was going to happen, he would have collided with me.

I see illustrations similar to this every day of my driving life. I am forced to the conclusion that **there is something about driving which reduces the average person's brain to water.** To be in the correct position when moving along the road is a sign of the skilled driver. Having considered position when stopping and when moving along, it is necessary to think a while about position when changing course. Driving, as I am, considerable distances every day, I have opportunities galore to observe the driving habits of others, and, in the matter of changing course there are very few drivers who observe the most elementary rules of safety and courtesy in this respect.

First let us consider the case of the man who is turning left. In a road wide enough for three lanes of traffic a car turning left in front of me should not impede my progress in any way, not even when there is another car coming in the opposite direction. A man wishing to turn left should keep in to the left. If he cannot make a left turn without performing a half circle out into the middle of the road first, then he has no right to be at the wheel of a car. So many motorists give a signal for a left turn which should mean that following traffic may immediately overtake on the outside. Woe betide the motorist who commences to take this action without keeping a wary eye on the man who is going to turn left, because, more often than not, he will pull a yard of two out to the right after signalling a left turn in order to get a wide sweep round the corner. This is dangerous driving. Once a left hand signal has been given a car should move only to the left and never to the right. Yet hundreds of motorists seem incapable of turning off a road at all unless they can make a wide sweep first. Why is this? Have these people absolutely no brains or intelligence at all? Have they never been told how to turn corners? Do they not realise that the moment they pull out to the right with their left hand signal flashing they are asking to be involved in an accident? Even if this accident is avoided by the vigilance of the driver behind, they are still guilty of sloppy inconsiderate driving and deserve the very worst that can happen to them. If you wish to become a competent driver please learn to position yourself correctly in the road for the action you intend to take, then signal your intention clearly, check that you are free to make the intended move because a signal does not entitle you to anything at all, finally move only as indicated by your signal.

In passing, this matter of signalling and just what a signal means will bear consideration. The fact that you signal that you wish to do something does not mean that you are free to do so and everybody else must stop to allow this. Not at all. Signalling is indicating what you wish to do — you must not do it until you can do so without causing any other motorist to swerve or brake violently. **A signal is not a passport.** You are not entitled to make any move on the road just because you wish to. First, you must take care of traffic near to you by signalling this traffic on or by giving warning over a fair distance that you are going to make a certain move. The person who flicks on the indicator and immediately turns off a main road or into a parking space is an anti-social, unskilled, and dangerous driver.

What has been said about left turns applies even more so to right turns. A very large number of accidents are caused by traffic turning right. If you wish to turn right pull over to the crown of the road, signal that you intend to turn right, turn when safe to do so. Do not drive until the last minute well over to the left, then signal a right turn, and pull out into a car just about to overtake you. You are in the wrong even if you have 1000 signals flashing that you are about to turn right. Do not stop in the road when waiting to make your turn, so that following traffic cannot pass on the inside of you while there is space to spare on your far side. Thousands of motorists have this irritating habit. Such drivers have absolutely no idea of how to drive and are the immediate cause of much frustration and no few accidents.

The skilled driver takes up the correct position for his next move smoothly and in good time; signals well ahead of his intention to turn; turns only when it is safe to do so. Turns correctly into the road he is entering, entering only in the half of the road on which he will proceed.

Position is essential to perfection in the art of driving. The unskilled driver is always finding himself in the wrong position for a smooth execution of his intentions.

9   STEERING

May I be permitted to re-emphasise that a car is a machine like a washing machine, a vacuum cleaner, a sewing machine, a lathe. Just a machine, capable within its limits. Having absolutely no will of its own. Responsive to the slightest wish of its manipulator. It will not start itself up or drive itself down to the local. It will not turn itself right when you wish it to turn left.

When a man involved in an accident says, "The car just went all over the place," he should say he made a mistake and induced certain reactions on the part of the car. The car behaved as it could be expected to behave under certain stimuli.

I well remember the metalwork master at one of the schools I attended. If a boy approached him with a broken drill and said, "Sir, the drill's broken," the master hit the roof and shouted, "Boy, you mean you broke the drill." This lesson was soon learned by us all.

Your car will respond to you. The steering is there to be used. Do not be afraid of it. It can only respond to you. It will not act on its own. Yet motorists in the main behave as though the steering was something to use only if one must. Many a man who has been involved in an accident could have steered out of it. I saw a motorist with twenty yards to spare still run into another because all he could think of doing was to stand on his brakes on a wet road.

Always be ready to steer out of trouble. Always be ready to steer round a bend if you have entered it a bit faster than intended. Heavy braking in the bend may be disastrous. Good steering plus gentle gear changing will often save the day.

I see people halting before a space big enough to take a bus through. These unskilled drivers can only keep going in a straight line. Once even a small amount of steering is necessary panic arises. I cannot understand this. Steering is for use to guide a car, just as brakes are for use to stop a car. The skilled driver makes use of his steering to save time on a journey and to get out of trouble when brakes alone will not suffice.

Steering can also get one into trouble. Steering on a perfectly normal surface can be pretty violent without producing any alarming results. On a road which is wet, or affected by ice or frost, this is not the case. In such circumstances it is necessary that the steering be handled carefully, delicately.

On roads which may be slippery the endeavour should be to drive in as straight a line as possible and to steer cautiously at all times. Many a driver comes unstuck on a slippery road because he does not adapt his driving to the conditions. Heavy or violent braking will probably bring disaster in such conditions, and this is generally realised. Often, however, the driver who realises this point about his brakes will swing round a bend and produce a slide which could have been avoided by a more gentle turn.

It is a step toward efficient driving to be in command of your steering. You must be able to steer out of trouble; reduce the amount of time wasted by being held up when a space in front of you is sufficient to drive through if only you knew how; understand the need for delicate handling of the steering under adverse road conditions.

Many a driver loses hours of driving time in a month because he is constantly using brakes even before quite gentle curves in the road, when all he needs is to steer round the curve. To revert to the consideration of position — if you are driving always in the correct position for your next action your steering becomes easier. The man who is in the right position when entering a bend, and who follows the right line through a bend, has less steering to do, and hardly any need for braking at all.

Master the art of position and use your steering wheel more than your brakes and your driving will improve in quality and speed. The better you drive the faster you will drive. A driver who has improved his technique will find that without increasing his maximum speed he will raise his miles per hour. This is efficiency in driving.

10   STOPPING

We have said a certain amount about stopping under the heading of position, but it may not hurt to dwell a little longer on this because while a great number of accidents occur between vehicles which are both moving at speed, a considerable number also occur when one or both vehicles are stopping.

Have you ever run into someone on the pavement because when walking close behind them they have suddenly stopped dead and you could not, or you were not looking.

This happens quite often on the roads. I have observed it several times. We should learn several things when we see one vehicle run into the back of another, especially under normal road conditions. Firstly, the vehicle in front should not stop suddenly without a signal unless the emergency is such that there is no time to signal. In that case the vehicle following should be moving at a speed and distance which enables the driver to pull up sharply also.

Avoid forgetfulness. When driving in front of other cars remember they are there. Use your mirror. It is not there for the purpose of adjusting your tie or your hair. The mirror is for use to ascertain the position behind you prior to any change in your driving pattern. No driver should ever start, turn, overtake or stop, without first looking in the mirror. The good driver is always glancing in the mirror. The good driver is never overtaken by another car he did not know was coming. Likewise the good driver never pulls out in the path of an oncoming car. If you are surprised by an overtaking car or you pull out in front of a car about to overtake you, then you have not yet learnt to drive competently or safely.

To return to the immediate point. If you are using your mirror, you will not stop dead in front of another car without signalling, and if your mirror indicates someone is coming up fast you will stop slowly after signalling. You will not be so interested in finding a parking space that you will forget the man behind. You will not suddenly brake, and dart into such a space. If you watch some drivers you will conclude they are playing a kind of automobile musical chairs, darting swiftly and surely to their parking spot as soon as the necessary spot appears. Damn the other man is the attitude, I'm entitled to park aren't I? The short answer is yes, if you consider other drivers while doing so, and no, if you do not.

When driving behind another car consider several things. Firstly, if it is clear that you are behind an unskilled driver, and this will be obvious enough, expect anything. Keep a reasonable distance and keep your eyes on the blighter. If you really do this you will not run into him whatever he does. Do not let your attention wander or the man in front may stop and you will not. In these circumstances you have only yourself to blame. Skilled driving is keeping out of accidents not merely avoiding causing them.

The above applies mainly to driving in crowded town areas. What about the open road. Let me tell you two stories.

The first is this. I once drove a Jaguar and the speed at which I could safely travel and safely stop was much higher than that of the average family car, and I would not drive an average family car in the same way at all. But other people do. On the A20 returning from Dover to London before the 70 mph maximum speed limit was imposed I passed a small Ford. I noticed in my mirror that the driver of the Ford immediately gave chase. I was touching 70 miles an hour and so was he very soon. I thought that this driver would get himself into trouble so I increased my speed to 90 miles an hour and over. The Ford was being outpaced. The roads were very dry and dusty. It began to rain quite heavily and the road to me seemed to be tending towards a dangerous condition. I therefore began to slow down to about 60 miles an hour because I could see a hold-up in the traffic about 200 yards ahead. The Ford was soon right up behind me. I thought to myself that this man would run into me as soon as I stopped, and we were nearing the obstruction ahead which he could see as well as me. I began giving the slow down signal, and then stopped in the line of traffic. The man behind, so excited at being able to keep up with a Jaguar now realised he was required to stop. He braked. His car went all over the road and crashed. His first words when I checked that he was all right were, "Bloody hell, I've only had this car two weeks, disc brakes on the front too, I braked and nothing happened."

It is a fact that this man, who, I discovered, held a quite responsible position in life, and was rated a top business executive, had no idea why he had crashed. It was all too much for him. Can you explain this to me? May I say it again — why do people, when entering a car driving seat, leave all their brains behind them? How can intelligent people become such absolute idiots? Accidents will continue until drivers realise that the application of intelligence is more important in driving a car than in their business life, for, after all, what use is a good business to a dead driver? To paraphrase the Scriptures, "What shall it profit a man if he reach the heights of business and social success, and lose his life in his car?" Through business success you can become rich enough to own three cars and be killed in any one of them. When people use their intelligence in their driving the accident rate will fall, and not before then.

The second story is similar. This time the scene is that morning stretch of motoring murder, the old A4 between Slough and London. Between two sets of lights which are about three quarters of a mile apart I stopped behind other traffic which had pulled up very sharply only to receive a nudge at the back. I got out and so did the driver behind me. He said, "We're lucky. Brakes just would not hold. Could have been a lot worse."

The distance from the last lights was about 400 yards. How can a man driving between sets of lights get himself into trouble after 400 yards, knowing the road he is on, and knowing he may have to stop for the next lights very shortly? How can a driver talk about luck? I do not believe in luck in driving, I believe in skill. The person who is lucky today may be unlucky tomorrow. The person who is skilful today will be skilful tomorrow.

What does a man mean when he says his brakes would not hold? This has no meaning for me at all. Does he mean he has to run into other cars whenever he wants to stop? Does he mean that he is committing a criminal offence driving a car in an unsafe condition? Does he mean he applied his brakes too late? Does he mean he was too near to my car? Does he mean he was thinking about something else, or looking at a pretty girl, so that when he knew he needed to stop he could not in time?

"Could have been a lot worse." Suppose that next time it is a lot worse, and he kills somebody! What will the police make of his "the brakes would not hold".

It is not fast driving which is unsafe, it is careless driving which causes the trouble. Every driver must learn to drive so that he can stop in the distance he has at the speed he is doing in the time available. If your reactions are very quick you need less time and less distance. If you have ABS brakes you need less distance. If you have not got ABS brakes and are behind someone who has you need more distance however quick your reactions are. I wonder if you as a motorist ever consider the driver in front of you. Do you note that, on a wet road, you are travelling pretty well all-out, behind a car with more speed than you and better brakes? Do you note this and react safety-wise from the observation? Do you allow any more stopping time on a wet road than on a dry road? Do you allow any more distance when behind a vehicle with better brakes? Do you? Or, would it be true to say that you never thought about the matter?

There are some ghastly accidents at night. Do you drive at night too fast to be able to pull up in the space equal to the distance you can see? If you do, what happens if a big lorry is pulling out of a layby as you come round a bend? The lorry is broadside on to you and you do not see it quickly as there are no lights on the side of the lorry. By the time you see it, can you stop?

In town, or in open country, when you intend to stop signal this intention and wave traffic on. Do not presume that the man behind knows what you are going to do. Signal in good time, not after you've stopped as many do. Remember to use your mirror and consider other road users. Never do anything without warning. Never be in a position where you cannot stop. When stopping in traffic stop in position. Do not hog the road. When driving on wet roads remember you need more time to stop. When driving at night drive within a speed which enables you to pull up in the distance you can see.

Failure to stop your car may end your life, or someone else's life.

11   OVERTAKING

According to police records, careless or dangerous overtaking is one of the major causes of serious accidents, and therefore deserves a considerable amount of thought by the driver who wishes to become skilled in the art of driving, and, in fact, by every driver who would stay alive.

I often find myself held up on the roads by another driver who wants to overtake, yet seems to find it almost impossible to do so. There are times when I feel I cannot wait any longer, and I flash this driver to indicate that I am coming through past both him and the vehicle he seems unable to pass. The sort of driver mentioned here is a ditherer, lacks confidence, and would be better to content himself and stay where he is, and not worry any more about overtaking.

However, once I pass such a driver he follows me. My action gives him confidence and he pulls out behind me only to find himself face to face with a car coming the other way. I had time to pass before this car reached me, and did so, but the road was not then clear for another vehicle to follow me. This illustrates the fact that if you lack the confidence to overtake, it is extremely foolish to follow someone else without first checking that there is time for you to overtake as well as the man in front of you. There might not be.

A habit I notice frequently among motorists wanting to overtake is that they drive right up to a big lorry in front of them so that they cannot see the road ahead at all, and heaven only knows if they would be able to stop if the lorry, which may have special brakes, did. These people have no idea of the correct technique for overtaking. Watch them. Every so often they have to pull out to look up the road to see if they can pass. Each time they pull out blind in this fashion they are lucky not to be involved in a collision. They never seem to realise this and repeat their suicidal outward pull over and over again. If the lorry driver was to signal them to pass, they are so close they would not notice the signal. All the time other traffic is being held up behind them. This sort of idiocy again illustrates my contention that once in a car the average person leaves every vestige of common sense or brains normally demonstrated absolutely out of the reckoning.

To overtake correctly keep at least two car lengths behind the vehicle in front if this vehicle is taller or wider than your own car. From this position you can pull out to look up the road without doing so blindly. Also you can pull in to the near side if there is a left hand bend ahead and look round the nearside of the vehicle in front to find out if it is safe to pass. So many drivers only check the offside of the vehicle in front to see if it is safe to pass. This means that if the road is mainly tending to the left they will never be able to see that it is safe to proceed, and once again the traffic behind them is held up because they have not applied their intelligence to their driving.

First lesson then for safe overtaking. Keep your distance from any vehicle ahead. Check if the road is clear from the nearside when vision ahead is not possible from the off-side. Be professional in your driving by learning all the possible ways of improving and speeding up your driving.

A further point in connection with overtaking needs to be mentioned. This is the necessity to check for safety behind as well as in front. Many a time when I am passing a line of vehicles a car will commence to pull out right in front of me. Cars are supplied with driving mirrors because these are deemed to be necessary. Why on earth do not people use these mirrors? Are there really so many who want to die or be injured for life, or who do not care if they kill or maim other people?

Before pulling out check behind and do not pull out if anything is approaching either near or fast from a little further off. The fact that you have signalled that you wish to pull out does not entitle you to do so. Signals are only expressions of a desire or intention to make a certain move. This move should only be made when it can be made safely beyond any shadow of a doubt. The fact that you have given a signal will not make an accident any less tragic.

We have said therefore that it is necessary to check behind and in front before overtaking. Yet, it must be said, that accidents occur to those who make this double check. Why? It is really very obvious. They have checked, and an impression has been recorded in their mind, and they have proceeded on this impression indicating it is safe to do so, but the impression was wrong. How to avoid this?

The way to avoid overtaking accidents is to **learn to judge speed as well as distance.** One without the other is as bad as neither. The first lesson to learn is not to judge other drivers by your own standards. You normally drive at say 45 to 60 miles an hour and think this is fast enough for anybody. When about to overtake, you see a vehicle in front or behind at a distance of 100 yards. You pull out to overtake, but wait, the vehicle you saw in your mirror or approaching you from ahead does not accept your ideas of top speeds, and is approaching you at 80 miles an hour. You will never make it. The reason is that you have paid attention to distance only and not to speed, and as a result, maybe the death rate on the roads is increased. You are mostly to blame. Once again, a little intelligence would have saved the day. If you cannot use your intelligence when driving then get off the road. There is no room for sloppy sentimentality in these matters. If you cannot drive intelligently you have no right to drive at all.

This matter of judging the speed of a vehicle approaching either from ahead or behind is brought home very vividly to me every time I travel on the M1. I can say quite definitely that if I maintained my speed each time a car pulled out in front of me, I would be involved in an accident at least once every two miles. That is how bad the standard of driving on the M1 is. It is only my own preparedness for the stupidity of other drivers that keeps me out of accidents. When I am approaching at maximum speed a car will pull out 100 yards in front of me when it is travelling at 45 miles an hour. I wish I could stop each driver who does this and ask him why he thought it safe to pull out. Or better still it is a pity I cannot maintain my speed and show him the result without damage to anyone.

If you pride yourself on your ability to grasp things quickly, and think that you have a reasonable level of intelligence, please will you apply it to your driving.

One more point about overtaking. Do it fast and do not cut back in too soon. I watch drivers trying to overtake at times and they are not travelling any faster than the car they are trying to pass. Why on earth do these drivers try to pass other cars if they have not even learned the rudiments of driving. What happens is a car has to slow down behind another. Then comes an opportunity to pass. Does the driver, who has reduced speed considerably, change down in gear? Not on your life. He labours past in top gear at a speed which calls for a lower gear.

This is unintelligent, sloppy, unskilled, and lazy driving. More often than not the driver has to fall back again, and the traffic behind is held up once again by the utter stupidity or laziness of another driver. For your own safety, when you overtake, overtake fast. As fast as you can. Be in the right gear to achieve this.

Having passed, do not cut in sharply, especially if you have just passed a long distance lorry. The drivers of this type of vehicle are among the best in the world and the most courteous, and they have quite a job to build up a decent cruising speed. Don't cause them to brake unnecessarily, especially on hills.

Make your driving intelligent and courteous.

12 DANGERS

The worst accidents are often totally unexpected, and occur because drivers forget that **yesterdays conditions may not apply today**.

The man who starts out very early to work and crosses the same roads every day for years sometimes becomes very careless in his attitude. This carelessness arises because familiarity breeds contempt. This driver has taken the same route early in the morning for years and hardly ever sees another car on a certain part of the route. After some weeks he does not halt any more at a certain road junction because in his experience nothing ever appears at this point at the time when he passes.

Suddenly, one day, something is coming where nothing came before. There is a crash which would not have occurred had the driver not become too sure that nothing could be on the road at a certain time.

The same thing is liable to happen late at night on well-known local routes. Drivers must learn not to take anything for granted. The only safe way to drive is correctly at all times. All halt signs should be respected whatever the time, and however well known the route.

13   SPEED LIMITS

There is much debate about speed limits these days. It seems that the main idea in the minds of officialdom is that speed limits will reduce accidents. I think this is so much poppycock and have no hesitation in saying so.

Some of the worst accidents that happen today are head-on collisions. I do not think that a 50 miles per hour speed limit will in any way reduce such accidents.

I suppose the idea is that it was a mistake to build roads for speed, and that the easiest way out now that the roads are there is to restrict the speeds permitted.

I remember the opening of the M1, and the astonishment expressed by the Minister of Transport because a Jaguar sped up the road at 120 miles per hour. **What was the M1 built for – prams?** Why build motorways if speed is to be regarded as the wicked fairy? Why not build nothing but one-way-streets too narrow to allow any overtaking, and reduce all traffic to the speed of the least competent. The only difficulty remaining then would be to find a way of stopping the drivers going to sleep and running into each other.

Blaming the increase of accidents on speed means that the real cause of accidents is apparently being shelved by the powers that be.

The cause of accidents is bad driving. The really fast driver usually is not a bad driver. It is a fact that as the art of driving is truly mastered, speed follows automatically. The good driver is a relaxed driver who goes through all the mechanical movements without conscious attention and therefore has his eyes concentrated on the road. He drives at speed a lot more safely than the more immature driver does at 40 miles per hour.

I am thoroughly sick of the criticism of speed. Why do we not criticise the creeping and crawling driver, who is never sure what he is going to do, seldom looks in his mirror, is hardly ever in the right position in the road for what he intends to do next. It is sloppy, untidy driving that causes hold-ups and impatience – and often accidents as a result of this impatience.

It might be worth a thought to consider how many accidents occur inside speed limits at lower speeds than 30 or 50 miles an hour. Also, it might be of interest to find out how many accidents outside the speed limit occur at under 50 miles an hour, as this is the speed limit it appears likely will be imposed more and more on main roads.

While on the subject, we could go one step further and find out how many accidents occur at over 50 miles an hour. It might just be discovered that these are the least numerous.

The answer to accidents is better driving by all.

As to speed limits. I believe that every speed limit in the country should be abolished immediately, whether this be a 30 or a 40 or a 50 mile limit.

Prosecutions for speeding should cease immediately. The police could then spend their time doing something to reduce the number of accidents on the road by correcting sloppy and rank bad driving. Instead of prosecutions for speeding there could be far more prosecutions for careless or dangerous driving. After all, a man driving through a deserted village at 50 miles per hour at 3 in the morning who is prosecuted for speeding is far less likely to be the cause of an accident than a driver who turns right from a line of traffic in the rush hour, without any warning even though travelling at only 30 miles an hour.

It seems to me that we have everything back to front today. Speed kills, we are told. It does nothing of the sort. People are killed by careless or dangerous driving; sometimes speed is involved and sometimes not. Speed, pure speed is not the cause of accidents. Speed limits do not reduce accidents. They never have, and they never will. Good drivers do not have accidents, and good drivers are apt to be fast drivers. One policeman who booked me for speeding in the early hours of the morning congratulated me on my driving, and said it was a pity he had to book me, but his sergeant had signalled him to do so. I wonder if the police really believe in the bogey of speed. Do they believe it when speeding themselves in the course of their duty?

I am prepared to stand by my point of view. I will wait and see the speed limits added here and there. The accidents will continue. **Only if driving standards improve will accidents decrease.** If someone will have the "guts" to scrap all speed limits, and instruct the police to watch for **bad** instead of fast driving we shall have taken a big step toward the reduction in the present rate of slaughter on the roads.

14 CONDITIONS

When considering safe driving it is important to consider the following conditions: the condition of yourself; the condition of your car; the condition of the roads; the condition of the weather.

Firstly then, let us consider one's own condition. The condition of mind and the physical condition of the driver of a car has some bearing on his proneness to accident.

As far as physical condition is concerned it is a fact that drivers have been killed or injured through falling asleep while driving. When very tired the driver will be well advised to stop and sleep for a short time, and not to continue driving once he finds himself nodding.

Another physical point of interest, even vital interest, is the question of sight and age. Many a man who has driven in his youth, and even into his forties at very high speeds, may find as he grows older that he cannot see as well in the dusk or at night. He should drive more slowly to allow for this failure in sight which is apt to come with age. This failure may be gradual, and it is essential that top speeds be reduced in accordance with this unfortunate decline.

With age the time needed for reaction to emergency may well be longer, and this is another reason why top speeds should be reduced. Speed is not dangerous to a skilled driver with all his faculties working to the full, and his reactions fast, but once there is any decline in the faculties, speed should be reduced because although the skill will still be there in the overall method of driving, the sight may not be as good, nor the reactions as fast. Be wise, as you grow older, take a little more care, and you will continue to live.

The next condition we must consider is partly a physical and partly a mental condition – drunkenness, and not only drunkenness but the effect of alcohol on a man or woman who could not be said to be drunk necessarily.

The question of drink and driving is a difficult one because all people are not affected in exactly the same way by the same amount of drink. I have often drunk a bottle of wine and then driven a couple of hundred miles in perfect safety, but I know people who could not do this.

One thing that drink does is this. It increases confidence. In a driver who is already very skilled this may not cause him to be in any danger at all. On the other hand to increase the confidence of a driver who, at best, is not very skilled, is likely to be fatal.

It is difficult to get people to be honest about their driving. I know people who think they are good drivers, but I would not let them drive me 100 yards, or at least, not a second time. These people, who already think they drive well, are likely to believe they are Grand Prix class after a few drinks. They are not drunk, but they are extremely dangerous.

Drink is referred to as dutch courage. A person will have a drink before facing a difficult interview or task in order to have more confidence. This feeling of confidence in an unskilled driver is a very potent danger, and here lies the biggest danger with drivers who drink and drive. A person who feels confident will act that way, and the fact that the person drives dangerously with absolute confidence is not going to reduce the severity of the accident, or save the lives of the innocent people involved.

I could never say, "If you drink, don't drive," but I would call for the greatest care in this matter. After all, if you drive under the influence of drink and kill somebody, you are a murderer. There is no doubt about this at all. A person who indulges in an activity which may result in someone's death is responsible for this death if it occurs. I do not feel sorry for such people. I feel far more sorry for the woman, who after years of ill-treatment, finally kills her torturer. I consider the drunk driver, or the driver under the influence to be an absolutely despicable and unsocial person. I consider that a person who causes the death of others while driving under the influence of drink, should be banned for at least 5 years, if not for life. Certainly any driver twice convicted for such an offence should be banned for life, if not imprisoned for life for manslaughter.

It is no good saying sorry when someone is dead. You cannot replace a husband to a grieving wife, and how much regret will repay for the loss of an only child. In any case, people should behave in moderation. Everyone knows how much drink they can take without being affected. It is less with some than others. A man who makes a pig of himself with drink is a despicable person anyway. If in addition he then causes death or injury to someone else, he should be locked up for a period of time in which to reflect on his stupidity and unsocial behaviour.

I have a very poor opinion of people who drink too much. After all, there are other things in life to do and to spend money on, and in view of the danger that a person under the influence of drink can be, I think no punishment too great for such persons if they cause harm to their fellow beings.

Coming now to conditions of the mind that affect driving, first there is the mind-wanderer. It is a fact beyond doubt that a driver needs to be single-minded. A driver should have his mind on his driving every minute he is at the wheel. On the road things happen fast, and to have to bring your mind back from wandering may mean all the difference in the reaction time of a driver. If you cannot keep your mind on the job of driving I would suggest you give up driving. Travel by bus or train.

Secondly, there is the worrier — the person with business or family worries whose mind is full of anything else but the art of good driving. Many an accident is caused because a driver had something on his mind which occupied his attention. The driver should leave his problems for those times when he is not at the wheel.

The good driver is a relaxed driver. The bustler is not a good driver. The impatient driver is a dangerous driver. Life today is too fast in many ways. Everyone is in a hurry. When I speak of the skill of the fast driver this does not include the bustler — the man who is in a hurry, who tries to drive fast without skill. This man is aggressive, not skilful.

To drive safely, study to be quiet of mind. Total absence of strain, a mind free of pre-occupations, relaxed and definite, leads to safe driving. The whole of your mental concentration is necessary to master the art of skill at the wheel of a car.

Leaving the question of one's own condition, we must consider the condition of the driver's vehicle. A good driver will do better with a faulty car than a less skilful driver, but nobody can truly be said to drive safely in a car which is not kept in good mechanical condition. The driver who is content to drive with brakes which are not properly adjusted and in tip-top condition is a criminal. He is putting the lives of other people at jeopardy. Brakes, steering, tyres, must all be in proper condition all the time. To drive a car which is mechanically unsound, or with worn tyres, is an offence against other people who have the right to expect that if a fellow-being takes a machine out on the roads he or she will make certain that this machine is mechanically sound and unlikely to cause injury to anyone else. If it can be proven that someone had an accident resulting in the injury or death of any other person as a result of driving a car in unroadworthy condition, and that the condition was known, this driver should be sent to prison for committing an act of criminal negligence.

Surely a good conscience and the safety of both yourself and others is worth the cost of regular service of your car, or the cost of new tyres. **If you cannot afford to maintain a car in good condition you cannot afford to drive a car.**

We must now give a little thought to the condition of roads and the number of cars and pedestrians on these roads at different times.

To drive safely a driver must pay some attention to the road surface. A change of road surface can mean an accident to the careless driver. A skilled driver reacts to changing road surfaces, and does not drive at the same speed or even in the same style on different surfaces. Especially is this important to remember when the roads are slippery.

It is also of importance to drive with the movement of children and workers in mind. A road which is safe at almost any speed at 11 a.m. may be a black spot at 4 p.m. with considerable danger at even 30 miles per hour. Drivers must bear in mind that the same road is a different proposition at different times.

Lastly, we must consider weather conditions. Firstly, rain. After a long dry spell the roads may be very dangerous during and soon after a fall of rain. Here again, a little intelligence will keep you safe from trouble. Just take it a little slower in these circumstances.

Snow and Ice. I could not say that I am happy driving under these conditions, but I have driven about 20000 miles in them and I have not yet been involved in trouble as a result. To drive under difficult conditions, patience and super-concentration are essential. I well remember driving from Leek to London at an average speed of only 15 miles per hour in the very terrible weather we had in 1946. I did not have chains on my wheels. By patient and thoughtful driving I completed the journey when many other drivers did not. I well remember a large car overtaking me when I was travelling through six foot high snow drifts on hard packed snow. At the next bend this driver went straight ahead through the hedge into a field. I had no sympathy at all for this gentleman. Idiots deserve to reap as they sow. If conditions are bad do not defy them. Drive according to the conditions. After all, it is better to arrive in eight hours than to crash trying to arrive in two hours. I am a fast driver, I love speed. I revel in a fast journey by car, learning more and more of control under speed. This is heaven. But I would not, under any circumstances, drive faster than can be regarded as safe under the prevailing conditions.

Fog. I have just spent three days driving in appalling conditions of fog and ice. There have been thousands of accidents, and thousands of cars abandoned. I am ready to admit that an accident may occur in fog which cannot be blamed on anyone, yet I feel that most of these accidents could be avoided. I mean the kind that are collisions. For instance, during the trip I have just completed I counted more than 60 cars or lorries driving without any lights at all. Others were travelling with only side lights and in many cases these were so weak as to be useless. Avoiding these inconsiderate and stupid drivers was a nerve-racking business. I wonder what kind of person can drive in a fog without lights. No wonder there are so many accidents.

Another matter with relation to driving in fog bears mention. Namely the advisability to keep windscreen wipers working to clear the windscreen and improve vision. With 20 to 30 yards visibility I have been held to 20 miles per hour behind a line of traffic led by an idiot who cannot see where he is going because his windscreen is all covered with grime and he has not enough intelligence to use his windscreen wipers. Such people should never be allowed to drive. If they lack such elementary intelligence heaven only knows what other stupidities they will commit with perhaps more serious results.

Everything we are considering in this book brings us back to the same point over and over again. Good driving is intelligent driving. Bad and dangerous driving is nothing more nor less than stupid, uninformed, inconsiderate driving.

15 SPEED

I suppose the question of speed is the most controversial of the present moment. Articles appear regularly on this subject, and letters are written to the newspapers constantly on this aspect of driving. There are people who believe all speed to be a dangerous weapon without regard to the standard of driving. Others, more moderate, feel that speed is not to be indulged in on roads in this country because of the amount of traffic, and so on. Most letters to the papers are from people who like to meander in their cars and resent the driver who has somewhere to go, and wants to get on with his journey. The very people who criticise the fast driver, can be heard in hotels, bemoaning the fact that they are completely exhausted after driving 200 miles. As this took them most of the day I am not surprised. Nothing is more tiring than a journey cooped up in a slow moving stream of traffic.

To the skilled fast driver 200 miles is a mere bagatelle. To him, driving is a pleasure. I travel very considerable distances in the course of my business, and my customers express

surprise that I find it enjoyable. That is the whole point.
Driving can be enjoyable.

Speed without skill is dangerous, but as the art of driving
is truly mastered, speed follows quite naturally. Speed without
mastery of the art of driving is suicide or murder and perhaps
this is why there is so much criticism of speed. Speed in the
hands of a good driver however is perfectly safe. The skilled
driver is a smooth driver. His speed is not a series of jerks, it
is a flow. He drives in the correct position in the road,
anticipates what is going to happen in front of him and moves
accordingly in advance. When he overtakes the act of
overtaking is a continuation of a driving intention and not a
sudden and uncertain manoeuvre.

It must be understood therefore, that when I say speed is
not dangerous in itself, I am thinking of drivers who have taken
the trouble to master their car; who have learned good driving
technique. A truly fast driver who has made fast driving an
achievement has added to the pleasure of life. There is little
that gives so much satisfaction as to be able, with perfect
safety, to handle a car to the limits of its reasonable
performance.

A word of warning may be needed at this point to the
effect that a reasonable speed in one car may be dangerous in
another. The art of good driving is in knowing your own ability
and the limits to which any particular car can be driven. The
person who drives all cars alike is a fool.

A mature driver finds it easier to drive fast than an
immature driver to drive slow. It is worth-while to learn to
drive really competently because speed will follow without
effort. Skilled driving is effortless like skilled skating. The
unskilled on skates perform in jerks. The professional master
exhibits a continual uninterrupted flow. With driving exactly the
same will be found to occur. Good driving is relaxed driving,
and much of it is automatic, leaving the driver free to
concentrate on the road, to anticipate, to act promptly and

definitely. Apart from anything else, the time saved by looking ahead and planning earlier will increase average speed. Skill in driving means continuity which means time and temper saved.

The skilled driver moving at speed is the safest driver on the road.

16   HOLD UPS

It is worthwhile to consider why there are more accidents on the holiday routes and when traffic is heavy. It is not sufficient to say that where there is more traffic there will be more accidents as this does not necessarily follow. When there are only two planes in the sky over a distance of many miles, they sometimes manage to collide, and the most terrible traffic accidents can occur on roads almost deserted. Conversely, when the sky is full of planes, collisions do not always occur. The most congested roads are often free of accidents.

Why then do we read of such terrible accidents from time to time in the holiday season? I think I know. I think you will understand the reason after a little thought.

Hold-ups lead to accidents.

This is because **hold-ups cause impatience**. Hold-ups are caused in the main by sloppy and unskilled driving. Driving will be safer, and accidents reduced when more people drive with skill. There is no short cut to the desired result. Limiting speed will not necessarily have the desired effect.

How do great lines of traffic build up? Must it be so? Could it be avoided so that all traffic moved at a reasonable rate on the holiday routes? If so, one of the main causes of holiday accidents would be greatly reduced.

On a road to the coast a lorry is making its necessarily slow way along a road. A driver in a car falls in right behind, cannot see to overtake, and, in any case, is nervous about doing so.

This driver, although unable or unwilling to overtake nevertheless drives right up close to the lorry in front so that if the next man wishes to pass he must pass two vehicles instead of one. He now joins the queue, right behind the other car so that the next driver, if he would pass, has to pass three vehicles. Eventually this situation becomes a two-mile queue, and although there are breaks in the traffic coming the other way, it is now almost impossible for anyone to overtake in safety. Some get impatient and there is a nasty accident. Surely all this could be avoided by the cars first approaching the lorry leaving room for overtaking vehicles to get in front of them if necessary, or is it the policy to take the attitude, "If I cannot pass, nobody else shall, or all the parking spaces on the front will have gone".

The same thing happens in another way. A whole line of unskilled drivers stick out toward the middle of the road so that nobody can pass them safely if traffic is coming the other way. Why do not these people keep over if they are not going to overtake themselves? You must have seen lines of traffic, all well out in the road, and nobody getting anywhere at all.

A lot of hold-ups could be avoided by a little consideration. This is a small price to pay to reduce the accident rate and the wear and tear of other drivers' nerves.

17   LORRIES

Following up the contention of the last chapter, it would pay the average driver to observe the driving technique of the Knights of the Road, the lorry drivers. When two or three of these "monsters" travel together they normally leave space between themselves and the lorry in front so that faster traffic can make progress at least a step at a time. I wish car drivers showed half the courtesy that these drivers do.

The long distance lorry driver will also take an interest in the traffic behind him in other ways. He will advise when it is safe to pass and signal another driver on accordingly, or signal him to wait if there is danger. Why must the average car driver, who has a heater in any case, drive with every window shut, and never give any help to anybody? Why is this? It is part of the art of good driving to take care of other drivers on the road and to assist their progress. Yet, how seldom any car driver ever makes the slightest effort in this direction.

I wonder how many overtaking accidents would be avoided if the motorist jogging along and holding up everybody else would only indicate whether it is safe for him to be passed? It adds to the pleasure and reduces the monotony of driving at a slow pace, to be busy assisting other and faster traffic on its way. The car driver can learn a great deal from the lorry driver as to position in the road, and courtesy to other road users.

I am quite often appalled to see a driver who has been helped on his way by a lorry driver's signal, just sail on without any acknowledgement of the courtesy extended to him. If it is said that lorry drivers are not as helpful as they used to be maybe the take-it-for-granted conceit of the many car drivers of today is mainly to blame. For myself, when younger, I used to drive by night for preference because there would be fewer cars on the road, and I could rely for fast progress on the courtesy and help of the long distance lorry driver.

18   BRAKES

I have already emphasised the need for good brakes. May I repeat again that the cost of properly maintaining your car is less than the cost of accidents caused partly by faulty brakes, or tyres.

The point I most want to make in this chapter is that there are times when brakes should not be used. I have driven behind drivers who keep their brakes on all the way round a bend in the road. Steering is apparently something they have never heard of, or else they feel that there is something magic about it. Brakes they understand. Put your foot on and the car reduces speed or stops. But you do not want to stop going round a corner — you want to travel round the corner.

The correct way to take a corner is slow in and fast out. Brakes can be applied to check speed just before entering a corner but should never be held on right round the corner. Check your speed so that you feel correct for the corner, and **steer** and **drive** round the corner. If you are entering the corner at the correct speed, and following the right line through the corner, you have no need of brakes. What's more, if you try this brakes-on-round-a-corner one day when the road is extra slippery you will wish you had not!

In wintery conditions, if you feel your car beginning to slip away from you, do not use brakes to correct this. You will probably only increase the spin or overturn. Here again, steering supplies the answer, but more will be said of this under the heading of skids.

Nowadays tyre bursts are not as frequent as they were, but if you should have a tyre burst, do not touch your brakes. To deal with a tyre burst steer while changing down in gear gently. Do not change down abruptly as this will have the same effect as heavy braking.

A great many drivers stop more often then necessary. Anticipation and forward planning will save you a lot of tyre wear, and will reduce the amount of braking you do in a day, thus increasing your overall average speed for a journey.

If you think ahead you will see things beginning to happen, or about to happen, and by lifting the foot from the accelerator you may find that the situation has changed by the time you reach the point at which you might have had to stop, and you can proceed without the need to halt at all. If you do not think ahead, you will arrive on situations unexpectedly and will have to brake much more often. Skill in a car is reflected in smooth driving.

19   GEARS

Today it is common to buy a car with automatic transmission. For city driving such a car may well be a blessing, but, in my opinion, for general long distance driving in all conditions, manual change is essential. I have had one car fitted with automatic transmission, and sold it in four months. When driving fast it is a blessing to have a gear change as well as a brake pedal in an emergency. In icy conditions the same applies.

I know there are people who say, "brakes for stopping; gears for acceleration." One police chief recently said gears should not be used for slowing down or stopping. I consider this is absolute nonsense. There are many times when the use of gears to check the speed of a car is most essential. Often in approaching a corner, the best possible technique is a slight braking, change down, and into the corner. On a slippery surface a gentle drop in gear is a better move than standing on the brakes.

It is a fact, beyond doubt, that a great many drivers do not like to have to change gear. I suppose they are lazy drivers. Also, they miss a great deal of the best driving if they neglect to use their gears to the full. Let me illustrate that, in fact, it is a feature of sloppy and unskilful driving, to neglect appropriate gear changes.

Firstly however, let us consider the time needed to change gear. Many people, when learning to drive, and some for a while after, seem much put out by gears. They could manage driving quite well if it were not for the gear changing. Changing gear is as simple as can be, but because it is one of the things not immediately grasped by many learners it becomes invested with a degree of difficulty which does not exist. When learning to do several things at once the learner usually makes the most mistakes with the gears. Drivers should understand that this initial difficulty can be thrown off immediately one has learned to drive. After all, changing gear should be an automatic matter, like tapping the space bar on a typewriter or wordprocessor. Gear changing and steering should be automatic so that a driver can keep his eyes on the road and his mind free for anticipation and correct positioning in the road. A driver should never need to consciously change gear.

Depressing the clutch pedal and moving the gear lever must become automatic and smooth if you are to be a good driver. There is no need to think of gear changing as being difficult. It can be made difficult; it can be made automatic and simplicity itself.

If you have found gear changing difficult, practise sitting in your car without the engine running, and change to first, second, third, fourth, fifth (if you have it), reverse, until you can manage to do this with your eyes shut. Keep practising until you will never look at your gear lever again when needing to change gear while driving.

Having mastered the idea that gear changing is the simplest thing in the world, take a further step. Realise that gear changing does not have to be a slow, laboured business. I normally change through three gears while most drivers are pulling off in first gear. The average driver behaves as though his gears are new laid eggs. This is not necessary.

When moving off from traffic lights many drivers are so slow that far less people get across the lights than should. I find that I am unable to start for several seconds if I am behind three other cars. If I do start, I have to slow again or I would be through the back of the car in front of me.

This is because valuable time is wasted by drivers who cannot get into gear and move off without making a ritual of the matter. Journeys take longer and time is wasted unnecessarily, and to a person on business this is most frustrating. When I am the first car to move over traffic lights I am usually 50 yards the other side before the next driver has moved off.

When moving off, engage first gear instantaneously, accelerate medium fast, within five yards change into second gear, accelerate hard, and change into third gear within ten more yards, and then on into top. This is the way to drive off from a standstill.

Gear changing can be slow or fast according to the outlook of the driver. Fast will not mean damage to the gears if you have learned to co-ordinate your foot pressure on the clutch with the gear lever manipulation.

There is absolutely no reason why some drivers in modern cars should hold everyone else up because they make a meal of gear change and take off. The professional touch in driving includes the ability to be fast through gear changes.

We have discussed the time to take in changing gear. We must turn our attention now to the occasions when one should change gear. Most drivers do not change gear often enough. All drivers should know the speed at which each gear is most effective and the engine not labouring.

Perhaps a few illustrations of the failure to change gear as and when necessary will help to elucidate this point. Have you seen a driver forced to slow down before overtaking. Does the driver change down in gear, ready for a powerful surge to overtake as soon as possible. Often he does not with the result that when he commences to overtake he is in too high a gear to get up a good speed, and is therefore forced to crawl past the vehicle he has to overtake. In this way he increases his own risk of accident and is a confounded nuisance to drivers behind him who also wish to overtake.

The same thing can be observed in drivers going up hills. I often have sailed past other similar cars to my own on hills because I am in the correct gear and they are not. A great many drivers go up hills in top gear despite the fact that their speed is dropping with every yard, and their engine is labouring.

If you want to drive safely and skilfully learn to drive in the correct gear for the effort you are making. Change gear going uphill; change gear before overtaking after slowing down; change gear always after a slow down for any reason. Be in the correct gear at all times.

20   SKIDS

The best thing to do about skids is to **avoid causing them**. Most of the slides that occur on the road are caused by the driver who suffers them. There are, of course, occasions when the roads are icy and a car will start to slide without any cause other than the road conditions themselves. Even in these conditions however, many slides that become quite dangerous could have been avoided, and if not avoided, they could be checked before the point of no return.

The most important lesson to learn in this matter however is how to avoid skids. Prevention is better than cure because most of the drivers who do not have the sense to avoid actions likely to induce a skid, are most unlikely to be able to correct the skid quickly. Their actions are more likely to intensify the slide that has begun.

Learn to avoid skids before learning how to get out of them.

Sloppy, inattentive driving leads to skids because such drivers are constantly surprised by what is happening around them. They have to brake sharply at the moment of surprise and they induce a skid which was totally unnecessary. Had they been paying attention to their driving, looking well ahead, anticipating the actions of other motorists, they would not have found themselves in the emergency that brought about the heavy braking that caused the car to slide.

The driver who knows how to drive very seldom finds his car in a skid, perhaps never at all.

Be advised. Drive with care and awareness on wet roads. Keep your foot off the brakes when in a bend. On ice or snow drive slowly. Use gears as well as brakes to slow down in certain circumstances. When using a gear change to slow down on a slippery road do it delicately not roughly.

Now, we must consider how to correct a skid. This is really quite a simple and straightforward action. If the back of your car is pulling away to the right steer to the right. If the back of your car is sliding to the left steer to the left. That is all you have to do to correct a back wheel skid. What you must not do is put your brakes on and lock the wheels. This will only put you into a rather spectacular spin.

How about front wheel skids? These are very rare, and you cannot be sure of finding your way out of such a skid. You must steer the best you can, change down gear, and apply the hand brake which acts only on the back wheels. The moment you feel the front wheels beginning to grip again, accelerate.

21   THE POLICE

I have already indicated that I think that the police are far too preoccupied with parking and speeding to be able to attend to prosecuting the many bad and careless drivers on the road. I have even seen two policeman just make a face of disgust after seeing a piece of bad driving. Nothing whatever was done about it.

While we have Ministers of Transport whose obsession is speed no doubt the police will have to continue to spend their time prosecuting people for speeding and doing next to nothing about the many flagrant cases of bad and careless driving that take place on the roads every day.

In the **Sunday Express** I read a letter from a reader which told the following story. Apparently an old lady practised obtaining help crossing and recrossing a street in order to meet people. Apparently this lady had been knocked down and taken to hospital twice in a year, yet she continued the habit. According to the letter writer a tolerant policeman watches this happening and does nothing about it.

Are the police just a little bit too tolerant of everything to do with the motorist except speed and parking? It seems to be a fact that if you park in the wrong spot you will be prosecuted. If you travel at 50 mph at three in the morning when there is no danger to a living soul you will still be booked for speeding if in a speed limit area. Yet, you can turn right without a signal right in front of the police and nothing will happen to you unless an accident occurs.

It seems to me that we have forgotten that prevention is better than cure except where speed is concerned. This would be all right if it were true that speed is the main cause of accidents. This is not so however. Accidents are caused by careless and unintelligent driving, and I want to suggest to the powers that be that if the police were instructed to take action against careless and dangerous driving and never against speed as such, the accident rate might be drastically reduced.

All prosecutions for speeding in the absence of careless or dangerous driving should cease at once. The police should be released from patrolling the roads looking for people driving at speeds over a certain limit. Instead of this, they should leave the good fast competent driver alone, and concentrate on stopping drivers whose driving is sloppy, careless, or dangerous.

The day that the anti-speed mentality changes to one of determination to wipe out unskilled driving we shall have started on the way to a reduction of the number of deaths on the road, and even bicycle-riding Ministers of Transport will be able to meander about the country in safety!

22   HOW TO AVOID PEDESTRIANS

The number of pedestrians knocked down on the roads each year is far greater than it need be. Pedestrians and cyclists sometimes do behave very foolishly on the roads, and certainly are much to blame for the accident rate as far as they are concerned. It is possible however for a driver to do much toward avoiding pedestrians. Anticipation will always reduce accidents.

Many of the accidents involving pedestrians take place in crowded streets with cars parked on both sides of the road. To avoid children, adults, dogs, and cats in such circumstances is difficult but not impossible. Most pedestrians and animals that are injured receive their injuries because they emerge suddenly from behind a parked vehicle into the path of an oncoming motorist.

The art of driving includes ways and means of avoiding all kinds of accidents including those to pedestrians behaving in a foolish manner. One way of being forewarned of pedestrians who may suddenly walk out in front of you is to keep a watch beneath the parked cars at the side of the road. In this way you can see the feet and legs of people before they step out. By keeping watch in this way I have been able to stop in readiness for someone walking or running out into my path.

Many a pedestrian would have been knocked down by cars driven by myself if I had not adopted this method of being forewarned of their approach.

If you will practise this method of keeping an eye open for what is likely to happen, and in addition, always be paying full attention to your driving, you will probably avoid the jay walkers and the unthinking pedestrian.

23   HOW TO AVOID ACCIDENTS

Reflecting, as I often do, on the number of accidents that occur, and especially on any that I have witnessed myself, I am becoming more and more sure of my ground in contending that accidents are caused by lack of intelligent driving, and by drivers whose attention is not concentrated 100 per cent on the job of driving.

The hallmark of the top driver is **anticipation**. Without anticipation a driver is like a ship without a rudder. The driver who cannot look ahead cannot avoid accidents and, in fact, will often be the cause of them.

To anticipate presupposes concentration on one thing only, namely driving. Any mind-wandering or looking away from the road means that the driver is not even concentrating on the present let alone anticipating the future.

If the great majority of drivers were accident avoiders instead of accident prone we should have moved a long way toward reducing road casualties considerably. The art of avoiding accidents is only a matter of intelligent anticipation.

No sane person doubts that prevention is better than cure, yet very few drivers seem to have learned how to anticipate trouble and avoid it on the road.

The proof of this statement is seen in the fact that the police are constantly informing us that the worst occasions for accidents are when traffic is turning right; at cross roads, and when overtaking. These are exactly the occasions when intelligence indicates that more accidents would be likely. In other words, at the very spots, and on the very occasions when accidents can be expected, they occur most.

This indicates that anticipation is not very much in evidence. Surely every motorist in the country with any grey matter at all could resolve to take more care on occasions and at points where accidents are known to occur.

Whenever approaching crossroads or even a side turning that I cannot see along, I am ready for a possible emergency. When travelling fast in the country I give a gentle toot on the horn whenever I am passing a concealed turning. This has resulted in a squeal of brakes from such turnings on several occasions. In other words, the anticipation of possible trouble which caused me to sound my horn, stopped someone from coming out into the main road in my path.

There are different views on the use of the horn. I have no doubts in the matter. I believe in using my horn every time I think the giving of a warning may prevent an accident. I would rather make a noise and avoid an accident than keep quiet and have a much noisier collision.

I have noticed drivers moving very fast along a main road apparently without a thought for the possibility of traffic coming out from a side road. Admittedly they have the right of way, but this is little consolation in the ambulance on the way to hospital. Drivers should always expect trouble from all possible sources and guard against it; be ready for it.

If you anticipate you turn what might have become an emergency into a safe passage. Emergencies come mainly to the thoughtless and to those whose attention is divided instead of being 100 per cent concentrated on the road.

You can often tell that another driver is going to pull out in front of you just as you yourself are going to overtake. By watching the movements of the head of the driver in front, you often are able to know what he will do next despite the fact that he does not signal his intention. If you collide with a man and say it is his fault because he turned off without signalling, you may well be right without being any more clever than he is. He was not being very smart to do what he did, but you were not very smart to run into him when you might well have anticipated what he would do next.

The skilled driver who has mastered the art of driving to a high degree has time to anticipate and therefore to avoid accidents in which others would be involved. It is therefore well worth your while to study to be as good a driver as you can. The more you learn to undertake the more mundane movements of driving automatically, the more you can concentrate on the road. The more you concentrate on the road the more you will learn to anticipate and the less likely you will be involved in accidents.

If you anticipate enough you will participate less in accidents.

24   THREE-LANE ROADS

Head-on collisions are the ones I understand least of all, because I think that a head-on crash must be avoidable always. However, these crashes do occur, and quite frequently on roads wide enough for three cars at the same time. In most areas centre lane markings have been removed or have been marked with double white lines on one side.

These crashes are usually very serious, and it is most important therefore to find ways and means of avoiding such smashes. Personally, I like three-lane roads least of all, especially in rush-hour traffic when the most glaring risks are taken by drivers anxious to get home while their dinner is still worth eating. Following a great deal of frustration getting out of a city in the evening a lot of drivers act just like lunatics once the road widens a little, and there appears to be a chance to get a move on at last. I like to move myself, and really fast, but only when I can do so safely. There is a world of difference between good, controlled, fast driving and the insane and aggressive fight for the road that one so often sees on the way into and out of big cities, morning and evening.

On three-lane roads our intelligence tells us that traffic can only proceed in one direction at a time in the centre, yet these appalling centre-lane crashes continue. When already in possession of the centre lane I have had drivers pull out from the opposite direction and drive straight at me. I suppose the idea is to frighten me into making way for these louts of the road. Quite often I have done so by some very difficult movement, because I do not intend to be involved in accidents. However, the point is that everybody does not react quickly, and often a crash results because the bully does not succeed in forcing the other man out of his path. This is a ridiculous state of affairs.

Rules for safe driving on three-lane roads are quite simple and extremely obvious. Let me emphasise a few tips which will keep you out of trouble on such highways.

If about to overtake into the centre lane, having observed that this is clear, flash your headlights on and off two or three times to indicate to anyone thinking of doing the same thing in the same or the opposite direction that you are pulling out.

Never pull out into the centre lane if there is a curve in the road ahead that is not more than 200 yards away from you, and not even then if you cannot see any spaces in the traffic ahead into which you can pull if you meet some idiot coming at you round the curve who has not observed this obvious precaution. It is no good saying in hospital that the other driver came round the corner blind. Of course he did, and this was criminally dangerous driving — but — you should have anticipated that just this might happen.

It is a golden rule of safe driving that all the possibilities of a situation should be considered by a driver before he overtakes on any road, and certainly on a three-lane road, and even more certainly on a three-lane road during a rush-hour period. Surely all this is plain common sense?

Lack of anticipation is the contributing factor to many an accident. That is why I say that accidents are caused, not by speed as such, but by unintelligent driving. The driver who anticipates, and drives in position in the road, is safe at any speed, anywhere. His speed will always be controlled by the demands on his intelligence made by the situation of the time and place.

One more tip. Be very careful about following another car out into the centre lane. The driver in front of you may have summed up the situation and decided that he has time to advance a short distance into a space ahead in the traffic. He may be driving a much more powerful car than you. There may not be time for two to overtake. Do not follow your leader — do only what you can do intelligently, and after seeing the situation for yourself.

In most areas three-lane roads have been abolished, but many roads carry three lanes of traffic when some are overtaking. The same lessons apply.

25   AWARENESS — ANTICIPATION

The reader will please forgive any repetition in this book because while the same mistakes continue to cause accidents I feel that there are many things that can bear to be said many times.

Awareness and Anticipation are the Safety Twins of all time. Where they are together in close harmony accidents do not occur.

By awareness I intend an attitude of mind when driving which does not allow anything to pass unnoticed that happens on the road. Awareness must precede anticipation, because if you do not notice what is happening at the time you will hardly anticipate what is likely to happen next.

When you are driving you should be noticing all that happens ahead of you, behind you, or on either side of you that may affect your driving safely. This is quite possible. I normally reckon to know if there is a car behind me, and whether it has an impatient driver or not, and at the same time, or within parts of a second, I have noticed two playful dogs on the left hand pavement, and a nervous little girl wanting to cross the road from the right, and have taken cognizance of the four schoolboys wobbling about on bicycles about 30 yards in front of me. Almost as soon as these facts have registered and anticipation has told me what may happen in all cases and to be ready for any one of these things, new impressions are entering my consciousness from behind, ahead, and both sides of me.

To achieve this kind of awareness, the mind must be concentrated absolutely on the task of driving. There is no room for anything else to monopolise the mind, although the radio may be on in the car, or a passenger may be talking to me, none of this takes my mind for one moment from concentrating on the road.

Because of this awareness, I can say, quite honestly, that there must be several people alive and uninjured today, who might otherwise have added to the number of road victims. Had I not been fully concentrating on my driving many people would have been involved in accidents with me. Perhaps, they would have deserved this. However, I am glad indeed, that, regardless of whose fault it would have been, I have been able to avoid accidents on the road. You can do the same, firstly by awareness and then by anticipation.

Anticipation will follow awareness quite naturally, as night follows day. If you fail to anticipate this is probably because you lack awareness of anything to anticipate.

Make up your mind that you will concentrate every second on the job of driving, and that under no circumstances will you allow your mind to be distracted. Learn to be automatic in the movements that do not need conscious thought; then concentrate your mind on the road. Observe, anticipate from what you observe, and drive safely as a result.

26   EYES

In many sports the secret of success is to keep your eye on the ball. A cricketer delights with a glorious innings of wonderful boundary hits, then lifts his head and is out. A golfer who does not keep his eye on the ball is doomed to failure.

Safe driving demands that a driver keep his eyes on the road. Any diverting of the attention from a continuous concentration on the job in hand can be fatal.

It is in this respect that drivers I observe on the road are the least careful. I would say that at least six out of every ten drivers lack concentration and do not keep their eyes on the road, and do not know what is going on around them. They often do not even know that there is a car behind them and that they are edging out to the middle of the road. They are astonished when a car they did not know was approaching, passes them. They write letters to the papers saying motorists should toot or flash their lights before passing instead of frightening the life out of them. They do not realise that this very request shows what completely unskilled and immature drivers they are.

The worst offence these motorist commit is in turning their heads to listen to or to talk to passengers beside or behind them. This is unforgivable, and in my opinion constitutes careless driving. We must make up our minds whether or not we wish to see a fall in the accident rate. If we do then all actions of a driver which take that driver's attention from the road are to be condemned.

When driving I carry on conversations with my passengers all the time if they so wish, but I do not take my eyes off the road for a single second. If my passengers cannot be satisfied to talk to me without my having to look at them, then I am afraid they will have to put up with not receiving satisfaction. There is absolutely no need why a driver of a car should take his attention from the road to look at a passenger.

If you feel that you may be regarded as impolite, explain to your passengers that you would rather look to your front and get them to their destination alive, than look at them and run into a tree or another car.

The other motorist that I am sometimes unfortunate enough to be behind is the gesticulating talker. Not only does he turn round to look at his passengers but he gesticulates as well, regardless of the fact that his car is veering towards another car and may be in collision with it.

Quite apart from the fact that the act of looking away from the road is dangerous in itself, as is taking the hands off the wheel to gesticulate, one should remember that in an emergency speed of action is necessary to save the day. If a driver first has to bring his eyes back to the road to appreciate what is happening before he can react to the situation, then this extra time may spell disaster. If he had to bring his hands back to the wheel as well, the situation is even worse.

All drivers who take their eyes from the road or their hands from the wheel in the manner mentioned above, are never likely to become good drivers, and, in fact, are to be numbered among the dangerous drivers on the road, and their lack of attention to their driving is more likely than not to be the cause of an accident. When it is, these people deserve to be punished severely. The rate of accidents must be reduced. To achieve this, careless and inattentive driving must cease.

27   STEPS TO TOP DRIVING ABILITY

If you believe that anything worth doing is worth doing well, and you apply this to your working life, and perhaps to your home life, you will have benefitted from this attitude. It is possible however, that, like so many others, you have not applied this outlook to your driving. May I suggest that this is a very serious oversight on your part. It is an oversight because you are missing the very great pleasure that comes from being master of a car on the roads. It is a serious oversight because if you drive without taking the trouble to drive well you are acting in an anti-social manner. To risk the lives of other people because you are too lazy or too stupid to become skilful in the art of driving is anti-social, and I feel that if we really wish to reduce accidents, we must begin to be honest in our attitudes to what causes these accidents.

I wish to suggest that to achieve top driving ability it is necessary to: drive regularly; drive considerately; drive cautiously; drive definitely; drive fast.

Firstly then, it is necessary to top driving ability to drive regularly. The Sunday and holiday driver is not likely to feature in the ranks of drivers of top ability because that driver does not drive enough. There is no reason why he should not drive safely by observing the rules of safe driving as these have been outlined in the foregoing pages, but to achieve the very top standard possible either on the roads or on the track requires habitual driving. With the habitual driver it is important that this acme of driving ability be attained because he is on the road more. Also it is the more unforgivable if this driver does not attain an efficient standard, because if he does not, this must be because he is either too lazy, too haphazard, too self-satisfied, or too stupid to do so. The man who drives every day and does not improve his standard, but continues to commit the same acts of carelessness or stupidity, is an enemy of society whether he likes to think so or not. If, because he is self satisfied and content with his driving he becomes careless on the roads through familiarity with driving, he is a fool to himself, and a danger to everyone else. You must have read of people who say when they have an accident that it is the first in 20 years. This is the more unforgivable. Twenty years of driving should make a driver as near perfect as possible, and every year of driving should make an accident a more remote possibility.

If you would become proficient, drive as much as you can. Learn all the time. Beware of the danger of self-satisfaction. Practise to do everything as perfectly as possible. Even in the absence of other traffic behave on the road as you would if the road was crowded. Always make your turns correctly, whether it is necessary or not. Do not learn bad habits because the road is clear, or you may repeat these when the road is not clear.

The second necessity to top driving ability is consideration for other road users. If you practise this you will not act dangerously or carelessly on the road. You will not change course without a signal. You will not brake and dart into a parking spot without any consideration for the driver behind you. You will not hog the middle of the road. You will not prevent another driver from passing you on the road, or accelerate like a jealous child with a toy when a driver is trying to overtake you.

Two people are necessary to a collision. If all drivers considered all other drivers on the road it would be impossible to find these two. The best drivers are the most courteous and considerate.

The next step to top driving ability is caution. The skilful driver is a cautious driver. He may not appear so at first glance as he speeds along the road, but he is. He moves faster and surer than the sloppy drivers around him, but he does not take uncalculated risks under any circumstances. His caution lies in his awareness and anticipation. He knows what is happening around him and what may develop. He is ready for it, whatever it may be. He does not only not cause accidents, he avoids becoming involved in them.

This caution does not mean dithering. The driver who dithers is over cautious and immature in his driving. The good driver is cautious but he moves resolutely and fast once he sees the road to follow. Because he is cautious he is not necessarily slow.

The caution which is an asset is positive rather than negative. The caution of the immature driver is shown by his inability to do things. The caution of the expert is positive and is shown by his ability to do the right things.

This leads naturally to the next point in the experts repertoire, namely the definite nature of his driving. Most bad driving is jerky, undecided, try-it-and-see-what-happens kind of driving. The expert is sure of himself. He knows his car. He knows himself. He observes the condition of the road and the weather. He takes up the right position in the road for the moves he wishes to make. He follows the right line through bends. He knows when there is room to pass; when it is safe to overtake. When he overtakes he moves fast and does not cut in. Everything he does is definite. While other drivers are thinking about a certain action he has taken it. Definite driving helps other drivers. If the man driving near you knows what he is doing it helps you to do what you wish without delay. Good driving never holds up anybody. Never causes anybody to swerve or brake unnecessarily.

If your driving is indefinite then it is not good.

Finally, top driving is fast driving. There is no doubt about this. You cannot be a top driver and not drive fast. The reason is simple. As you learn to handle a car, and most of your driving apart from road sense becomes automatic, you speed up without any extra effort. Slow driving is the driving of the unskilled. With them everything is slow because they have not mastered technique. They look at the gear lever in order to change gear. They have to look at the speedometer to know the speed they are doing. They have to brake at the sight of a bend in the road, however slight, and whether necessary or not. Everything about the bad driver is either slow or dangerous, or both.

The competent driver can change gear while the average driver is depressing the clutch. The expert will be in top gear before the immature driver has made one change. This driver is automatically faster. Skill in driving means fast driving. More than that, it means safe fast driving. Speed in the hands of an expert is safe. The expert will be always in the right gear and always in the right position in the road. All this means the expert driver will be faster without any extra effort.

If you would become a skilled driver pay heed to the need for courtesy, habitual driving, caution, definite driving. Speed will come automatically and safely.

28   CONCLUSION

This book has had one main aim — to try and show that accidents need not happen. That the number of cars on the road need not increase the number of accidents on the road.

This book has made these contentions — that accidents are caused by lack of intelligence in driving. That it is a mistake to make speed the main cause. That the police should concentrate on taking action against careless and dangerous driving and leave the good fast driver alone. That only when we cease being sloppy and sentimental in our attitude to bad drivers will we begin to reduce accidents by any appreciable amount. That any driver who will put their mind to it can become a really efficient and safe driver. That those who will not or cannot learn to drive correctly should be banned from driving. That those who drive to the danger of others are anti-social and merit punishment of some severity.

The author has enjoyed, safely, more than a million miles of motoring. Enjoyed is the operative word. Driving is great fun. Driving is an adventure. Driving is a challenge.

This challenge is worth taking up. The safety and happiness of others depends on the skill of today's drivers on the roads. Why not take up the challenge now? Decide before you put this book down that you are going to be a top level driver; that you will master the art of driving.

If enough people will respond to this appeal, the roads will be safer, and there will be less suffering and bereavement which could have been avoided. **You can be a skilled driver if you wish to be.** You can be responsible for much death and misery if you are too lazy or too stupid to master the car you drive. The choice is yours.